THE NEW FACE
OF NUCLEAR ENERGY

First published in French under the title: *Le nucléaire nouvelle génération. L'énergie qui réconcilie croissance et environnement*
© 2024, Hermann Éditeurs. All rights reserved.

Cover Design: Serge Lauret

www.editions-hermann.fr

ISBN : 9791037040305

© 2024, Hermann Éditeurs, 6 rue Labrouste, 75015 Paris.

Any reproduction or representation of this book, complete or partial, made without the explicit consent of the publisher, is illegal and constitutes an infringement of copyright. Any use is strictly limited to private use or quotation as governed by French Law (loi du 11 mars 1957).

JEAN-LUC ALEXANDRE

THE NEW FACE OF NUCLEAR ENERGY

Energy that reconciles growth and the environment

*Translated from the French
by Laurie Hurwitz*

TABLE OF CONTENTS

Preface 7

I. THE CURRENT ENERGY IMPASSE

1. A SUICIDAL PATH 15
2. CHANGING COURSE AS SOON AS POSSIBLE 21
3. DEGROWTH: A MISGUIDED SOLUTION 29

II. A COMPLETELY DIFFERENT APPROACH

4. GETTING BACK ON TRACK 37
5. THE NEW ORDER OF THINGS 45
6. A PIONEERING SPIRIT 53

III. RETHINKING THE FUTURE

7. THE INTERNATIONAL COMPETITION 69
8. A SYSTEMIC EVOLUTION 73
9. A CHANGING WORLD 81

Why I wrote this book 87

Preface

Our planet is overheating. More frequent and intense heat waves, more extreme precipitation, new types of droughts, the melting cryosphere[1], changes in species behavior—we are seeing these phenomena to different degrees in many different places. These are no longer millenarian fears but factual observations, so it should come as no surprise that this situation is giving rise to eco-anxiety. "The era of global boiling has arrived," declared UN Secretary-General António Guterres at the end of July 2023, the hottest month ever recorded in human history. In March of that year, the IPCC's final installment of the Sixth Assessment Report[2] warned once again of the increased risks facing our planet—and ourselves—as temperatures rise at a rate that has not slowed since 2014, the date of the IPCC's last full report. Experts predict that the rise in the world's average surface temperature is likely to reach 1.5°C by early 2030. Keeping this below 2°C would currently mean achieving net-zero CO_2 emissions globally, as well as other greenhouse gases (water vapor, methane, nitrous oxide, carbon tetrafluoride...).

In short, it wouldn't be an exaggeration to say that we're heading for disaster. If we don't find a way to change course in the face of this climate emergency,

---

[1]. All frozen parts of the planet, including ice, snow and frozen ground (permafrost).

[2]. The IPCC, or Intergovernmental Panel on Climate Change, published its synthesis report for the sixth assessment cycle on March 20, 2023.

our planet will become uninhabitable for many species, including our own. The term *energy transition* aptly sums up the new global imperative: the question of energy is at the heart of all these issues.

Energy is life

Before we have a look at the potential solution, one that has progressed beyond the research phase and will soon be available on the market, it's essential to understand the pivotal role energy plays in nearly all aspects of life.

The word derives from the ancient Greek term ἐνέργειᾰ, or *energeia*, which translates as "power in action." It refers to the ability to power the smallest gesture, whether it's picking an apple or launching a rocket, and it was at work long before the word was even formulated. Eve had to take action in order to get a piece of fruit, even a forbidden one!

The primary definitions of the word energeia are "force," "power" and "strength" and include the moral strength that enables us to achieve a goal. In addition to human strength, which for a long time was provided by slaves, we expanded our energy capacity by domesticating animals, then inventing tools, instruments and machines whose complexity has steadily increased over time.

Mankind has never ceased developing new devices to increase the power of energy since it was first discovered.

Preface

This concept goes far beyond the boundaries of industry or sector[3]. Energy is not merely a transversal and universal concept; its absence means the end of any real hope of achieving an ambitious goal. With it, nothing is impossible. Without it, nothing is conceivable. It's an Archimedean point. The real question is which energy to use.

Since the beginning of time, it's always been necessary to expend energy to achieve the slightest result. From animal power to nuclear power, we have transformed our relationship with nature. At first dominated by it, dependent on it and subject to its whims, we have gradually turned this relationship to our advantage: understanding the laws of gravity, for example, enabled us to fly and then to explore interplanetary space. It's understandable, then, that humanity has risen to the challenge of energy production with little regard for the consequences. But at least since the First Industrial Revolution, this frantic race has had disastrous consequences. If we don't properly manage the waste generated by energy production, we run the risk of producing pollutants that are harmful to our health, the environment and life in general.

Depending on the energy used, we can turn our planet into a dustbin or respect its biological and climatic equilibrium. This is the global challenge of the 21st century.

---

[3]. Companies that supply electricity, gaseous fuels, steam and air conditioning (there are about 30,000 in France, according to INSEE, the country's National Institute of Statistics and Economic Studies). The reason is simple: Every single act of our daily lives requires the use of energy. Not only does every country have an industrial sector dedicated to energy, but it pervades all human activities.

The 'energy wall': economic growth or environmental degradation?

Should we sacrifice economic development to protect our natural environment? Can we continue to benefit from continued growth without causing irreparable damage to our environment? Can economic progress be reconciled with respect for nature?

The answers to these questions form the basis of this book. Faced with this dilemma, we urgently need to reevaluate and adjust our behavior, even if we're tempted to give up.

If the only available solutions were to complain or abandon the idea of abundance once and for all, this book would never have seen the light of day. I wanted to challenge those who claim that we're forced to choose between abundance and caring for the environment, for three reasons of varying importance.

First of all, it's possible to build a well-tempered form of capitalism, in particular through the notion of "sobriety." This idea is gradually gaining ground. To live better, you don't have to spend like crazy—especially on energy—without considering the consequences of waste and the problems it creates.

Second, proponents of degrowth tend to forget that the world is not just made up of rich nations. Many populations are striving for better living conditions, if not for survival. Why should declinists have the right to impose a self-centered vision that ignores countries with rapidly growing populations?

Third, it's no longer feasible to go backwards; it's time to find solutions that work for the whole world based on pragmatic approaches, starting with the fundamental issue—energy.

That's why this book proposes a new solution: *using existing radioactive waste to produce nuclear energy that is non-intermittent, abundant, safe, inexpensive, accessible to all—and sustainable.*

Is it too good to be true? The following pages propose a three-step response:

1. Understand the determinants of the current situation. How can we characterize it? How can we develop a long-term approach? How can we avoid submitting to diktat?

2. Map out the path we need to take to avoid getting stuck in a dead end. What resources do we have? Is it possible to find new ones? Why should we believe it's possible to eliminate long-lived waste?

3. Consider the consequences of this approach. How can we get started on this new path? What changes does it entail? In what time frame?

I. THE CURRENT ENERGY IMPASSE

1. A Suicidal Path

Left to its own devices, nature does not produce waste; it recycles, recovers and reuses all of its resources. Each of us can easily replicate the practices of our distant ancestors: pick a fruit, eat it and then dispose of the residue. Three essential actions are taking place here. One, you ingest the food; two, your body's metabolism converts it; three, your natural waste products are absorbed by other life forms. Nature completes the cycle.

The verbs *take, make, dispose* therefore sum up all our activity on earth. Over time, these three stages have been developed and refined, allowing us to define our existence on the planet in more general terms.

The first (*take*) makes it possible for us to feed ourselves (plants, fruits, vegetables, meat, then various crops and minerals...).

The second (*make*) determines the way we adapt to our environment (body growth, reproduction, habitat, a wide variety of food, clothing, weapons and defense systems) in order to overcome the limitations imposed by distance (transportation, networks). It also facilitates the formation of multiple relationships and the provision of services among the millions (or billions) of people who are geographically dispersed.

The third (*dispose*) is the result of the previous two and leads to their renewal.

Everything might have been perfect if we'd stuck to the simple circularity of the Neolithic period, but economic development, while of great importance throughout human history, has brought catastrophic change, creating waste that is *non-recyclable* rather than easily assimilated.

The exploitation of the land and subsoil has not only made it possible to satisfy our ever-increasing needs and even to generate a surplus, but it has also led to the generation of a significant amount of waste, which has increased in intensity over the centuries. So much so that the *take-make-dispose* triad reached a turning point during the Industrial Revolution in Great Britain (1770-1830), and taking (especially the extraction of mineral resources), transforming (industry and services) and discarding (pollution) underwent a period of unprecedented and exponential expansion—this is commonly referred to as *economic growth*.

Since then, each of the three stages of this ancestral triad has become much more pronounced. Extraction has led to the gradual depletion of certain resources (fossil fuels, rare earths...); transformation has led to the almost infinite development of goods through their diversification (from soap to airplanes); emissions in the atmosphere (water vapor and CO_2) and waste on the ground (non-degradable substances such as plastics) have increased.

The current situation can be described as follows: While nature does not produce non-recyclable waste, modern man is accumulating more and more of it. The third part of this ternary model thus introduces a rupture in the history of the planet—the phenomenon that threatens us today is the clear consequence of our own actions, and all the more so as our actions increase[1].

---

[1]. The awareness of the risks to our planet had already inspired warnings from the Club of Rome, even if they did not go into detail about some of

1. A Suicidal Path

The search for energy has enabled human beings to build sophisticated civilizations, raise the standard of living, feed populations of several billion people (although inadequately for a good many of them), diversify consumer goods, walk on the moon and explore the universe. The search for energy has also created soil, subsoil and atmosphere that nature has been unable to absorb or correct. The verdict is in on this race to the finish line: Continuing on this path has become suicidal.

It's all because the energy forces at work have created a paradoxical situation. We have to keep increasing the use of energy to maintain the standard of living of an ever-growing world population while at the same time, this growth increases the burden we place on ourselves and the planet every day. In other words, we're victims of a scissors effect (declining revenues and rising costs) that makes it impossible for us to continue on our current path as if nothing had happened.

As we now know, stopping climate change requires the condemnation of fossil fuels (coal, gas, oil), but contrary to popular belief, coal is still the world's leading fuel, despite the huge investments made in renewable energy between 2010 and 2020[2]. This is simply because the majority of the world's population does not have access to the most efficient energy sources that do not emit carbon dioxide (CO_2). This gas, which is very harmful

---

our most harmful practices. In the 1960s, the Club warned of the "limits to growth," pointing out that the planet does not have an infinite supply of raw materials and that overexploitation would inevitably lead to their depletion, especially the non-renewable fossil resources such as coal and gas on which the industrial world was built.

2. 2,524 billion dollars were invested in renewable energies, while their share of global primary energy rose from 8.8% to 11.8%, an increase of only 3 percentage points. Source: https://ourworldindata.org/renewable-energy.

to the atmosphere, is one of the main causes of climate change, along with water vapor.

That's why a simple idea is beginning to take hold: first reduce and then, if possible, eliminate all polluting energy sources. We need to stop producing them, so we need to start decarbonizing the industries that emit them—in other words, power plants that make heat from gas. But that is much harder. How do we get there?

Electricity has been presented as a miracle solution. The problem is that we have to produce it. It's true that renewable energies such as wind and solar power are a step forward. They're complementary, but they won't be able to meet the world's energy needs. Not only because they produce energy intermittently (lack of wind and sunlight limits their production), but also because they're difficult to integrate into networks (transportation and distribution), both from an environmental and a financial point of view.

Despite its many benefits, nuclear power is often perceived as dangerous. In fact, it's a lot riskier to smoke a pack of cigarettes a day than it is to live near a nuclear power plant, but the prejudices against nuclear power persist. A series of accidents such as those at Three Mile Island, Chernobyl and Fukushima have left their mark on people's minds. This has led to a tightening of security measures, which have become drastic, without slowing down a certain amount of anti-nuclear propaganda.

But unlike that produced by wind turbines or photo-voltaic panels, the electricity produced by nuclear power is not intermittent. Its carbon footprint is virtually zero, unlike that of coal-fired power, and its cost is competitive.

Its biggest drawback is that today's nuclear reactors, known as pressurized water reactors (PWRs), produce very long-lived radioactive waste.

Whichever way you turn, the future seems shrouded in uncertainty to a degree that is demoralizing. There's no denying that our way of life is being tested. Climate change is becoming more and more apparent in our daily lives, and worry, a precursor to anxiety, is gaining ground in many people's minds, especially among young people. This isn't just the case in the most advanced countries; it affects more or less everyone in more or less all parts of the world, regardless of local socioeconomic inequalities. As a result, the very foundations of our world are now under serious threat.

What can we do?

2. Changing Course as Soon as Possible

Awareness of an increasingly untenable situation has now spread to all segments of the population in almost every country. This has led to the publication of the imperatives of sustainable development under the auspices of the United Nations to ensure the long-term survival of our planet. The United Nations Agenda 2030-2050 sets out 17 goals that member states committed to in 2015. These goals serve to highlight the necessity for a change of direction, and the direction is clear:

> "The Sustainable Development Goals are the blueprint to achieve a better and more sustainable future for all. They address the global challenges we face, including those related to poverty, inequality, climate change, environmental degradation, peace and justice. The 17 Goals are all interconnected, and in order to leave no one behind, it is important that we achieve them all by 2030."

This appeal from the United Nations is both a warning and an invitation to take immediate action, which means to:

1. end poverty in all its forms, everywhere

2. end hunger, achieve food security and improved nutrition and promote sustainable agriculture

3. ensure healthy lives and promote well-being for all at all ages

4. ensure inclusive and equitable quality education and promote lifelong learning opportunities for all

5. achieve gender equality and empower all women and girls

6. ensure availability and sustainable management of water and sanitation for all

7. ensure access to affordable, reliable, sustainable and modern energy for all

8. promote sustained, inclusive and sustainable economic growth, full and productive employment and decent work for all

9. build resilient infrastructure, promote inclusive and sustainable industrialization and foster innovation

10. reduce inequality within and among countries

11. make cities and human settlements inclusive, safe, resilient and sustainable

12. ensure sustainable consumption and production patterns

13. take urgent action to combat climate change and its impacts

14. conserve and sustainably use the oceans, seas and marine resources for sustainable development

15. protect, restore and promote sustainable use of terrestrial ecosystems, sustainably manage forests, combat desertification, and halt and reverse land degradation and halt biodiversity loss

16. promote peaceful and inclusive societies for sustainable development, provide access to justice for all and build effective, accountable and inclusive institutions at all levels

17. strengthen the means of implementation and revitalize the Global Partnership for Sustainable Development

2. Changing Course as Soon as Possible

Source: Cop 21 - 2015.

Anyone who considers these points carefully, one by one, can only be convinced of their validity. At the same time, they may be discouraging, or they may cause eco-anxiety because they don't offer concrete solutions. Promising the moon is one thing; achieving it is another. It is clear that these goals cannot be questioned, even if our ability to achieve them is debatable. Each of them is designed to promote the well-being and improvement of our existence, with the understanding that the planet is a common good. Therefore, it is important that they be written down and engraved in the minds of all citizens, from the general public to business leaders, actors and policymakers. It seems, however, that we are far from collectively embracing these aspirations.

A closer look at each of these goals reveals that they all depend on the seventh goal, to "ensure access to affordable, reliable, sustainable and modern energy for all."

This may sound dubious at first. How can goals number 5 (achieve gender equality) or number 10 (reduce inequality), for example, depend on universal access to energy? The answer is simple: To achieve these goals, you need data structures, trained professionals, education systems with computers, regulatory and human resources to punish violations, communication tools such as cell phones, and so on. All these tools require *energy*. Without them, the goal remains a wish, and a pious one at that. And there's more.

Goal number 7 not only serves to fulfill the conditions set by the 16 others, it also raises an important question of its own: What do the adjectives "clean" and "affordable" mean?

The first presupposes that the energy used, whatever its quantity, doesn't harm nature. It would be a great first if an energy source could both protect the environment and satisfy the need for cleanliness.

The second adjective is easy to understand: Energy must be available to everyone, at all times, without notice and at an attractive price. Otherwise, goals 1, 2, 3, 4, 8, 11, 12, 13, 14, 15 and 16 would have to be abandoned altogether!

2. Changing Course as Soon as Possible

In short:

This is the challenge we all face, and it's enormous.

Let's not forget that fossil fuels are becoming increasingly scarce (experts speak of "peaks," after which these resources will be irreversibly depleted)[1]. At the time of writing (summer 2023), they still account for 80% of the world's electricity, mainly because of their massive use in Asia, which has caused major problems (the greenhouse effect, threats to biodiversity...).

Mentioning the 17 goals of the United Nations, and in particular number 7, which is crucial, is not a literary exercise; it's a way of sounding an alarm. If we don't act as soon as possible, we will not only destroy our natural

---

[1]. As seen in a 2021 study by Jean Lahérrère, a French engineer, and reported by Philippe Gauthier, an economic and scientific journalist specializing in energy issues: aluminum (2035-2050), chromium (2019), cobalt (2030), iron (2030), lithium (2060), nickel (2025), rare earths (2110), uranium (1981), tungsten (2022).

environment, but also enter a period of social and political chaos. This invites us to meditate on the François-René de Chateaubriand's age-old reflection:

> "The excessive disproportion of wealth and living conditions was accepted while it was implicit; but as soon as that disproportion was generally perceived, the old order received its death-blow.... try to convince the poor, when they have been taught to read and no longer believe, once they are as well-educated as you, try to persuade them then that they must submit to every kind of privation, while their neighbors possess a thousand time their needs: as a last recourse you will have to kill them[2]."

The great writer's appalling premonition was not a figment of his imagination. If advanced countries reduce their activities, how can we justify the fact that the poverty of those who are most disadvantaged will increase?

And yet there is growing hope for universal awareness. This is a slow process, to be sure, but it's been gaining ground since the early efforts of the Club of Rome. The work of the IPCC, the compromise between developing and developed countries at the Rio Conference in 1992, the Earth Summit more than 30 years ago, the mobilization of citizens and the role of young people (such as Greta Thunberg)—all have contributed to changing attitudes. Political statements also bear witness to this, such as the op-ed published in the French daily *Le Monde* on June 21, 2023, on the occasion of the Paris Summit,

---

2. François-René de Chateaubriand, *Mémoires d'Outre-Tombe* [Memoirs from Beyond the Grave], Book 44, Chapter III, translated by A. S. Kline. See: <https://www.poetryintranslation.com/PITBR/Chateaubriand/ChateaubriandMemoirsBookXLII.php>.

signed by Joe Biden (USA), Ursula von der Leyen (EU), Lula (Brazil), Emmanuel Macron (France), Macky Sall (Senegal) and Olaf Scholz (Germany), among others, calling for a change of course as soon as possible[3].

Will action follow rhetoric? Let's hope so.

---

[3]. See: <https://www.lemonde.fr/en/international/article/2023/06/21/biden-macron-lula-we-must-prioritize-a-just-and-inclusive-transition_6034584_4.html>.

3. Degrowth: A Misguided Solution

Based on the above considerations and on the UN's effort to build a better world, there seems to be an obvious solution: slow down growth to the point of degrowth. To put it simply, it's not just a matter of reducing consumption and then reducing productive activity, but of reducing both at the same time.

In other words, both demand and supply will have to change. We need to eliminate extravagance and waste, and embrace sobriety—which the IEA[1] sees as one of the driving forces behind the energy transition—whether we're forced to (because of a collapse in purchasing power) or want to (because of greater awareness of the global interest). Changing consumer behavior to become true global citizens can have a real impact on energy production models, as evidenced by France's 9-10% reduction in electricity consumption in the winter of 2022-2023, or Europe's plummeting gas consumption since the war in Ukraine. But how can we convince everyone on the planet, regardless of their living conditions, to give up on improving them once and for all? Sobriety is undoubtedly a virtue, but betting on our virtuosity is a very uncertain proposition. Defending the benefits of sobriety is one thing; making it the *nec plus ultra* is quite another. While some people may be motivated to contribute to the global effort to save the planet, others

---

[1]. International Energy Agency.

would engage in unethical practices, such as cheating, to further their own interests. And game theory tells us that cheaters always win until they get caught, proving Chateaubriand right and condemning us to the worst crisis of all, leading to economic decline combined with the increased likelihood of political and social chaos.

Demanding sobriety may delay the looming disaster, but it can't eliminate it or postpone it indefinitely. Responsible companies have been integrating sobriety into their management for years by reducing the carbon impact of many of their industrial processes. That's why, without underestimating the indispensable joint efforts of consumers and producers, this book focuses on the key element of energy supply. As Jean-Marc Jancovici pointed out in 2022, "the part of industrial civilization that we will be able to save in the future will essentially depend on the proportion of nuclear energy that we manage to preserve[2]."

The logic developed by proponents of degrowth has all the trappings of evidence. Their argument is simple: In the *take/make/dispose* triptych, it has become vital to reduce the first and third parts. By doing so, we'd conserve remaining resources, stop dumping toxic products (especially plastics) onto the planet and stop damaging the atmosphere. The most immediate consequence would be the reduction of the second phase, making or transforming, which is at the heart of producing wealth.

So should we give up our long-held hope of improving our standard of living?

---

[2]. Interview in *La Nouvelle République*, May 28, 2022.

3. Degrowth: A Misguided Solution

The slogan "Save the Planet" is beginning to take hold, and some people believe that this is our only solution; they advocate the reduction of production and consumption in order to reduce emissions into the atmosphere and waste on the ground. In a word, to avoid catastrophe, they advocate an end to growth.

Could *degrowth* be our new horizon?

Advocates of this position have rightly pointed out that accelerated growth is a kind of parenthesis in history. What we call *stagnation* has been going on for thousands of years. It's impossible to calculate, even roughly, the rate of increase of production and trade over the last ten centuries due to the lack of statistics and precise historical studies. The first national accounts were not compiled until the 20th century; we can only proceed by what statisticians call "retrospective interpolation." It's conceivable that before the Industrial Revolution, annual growth rates were between 0% and 1%. This rate increased in the 19th century without ever reaching the level seen after the Second World War. Between 1945 and 1975, growth in GDP[3] (Gross Domestic Product), the internationally accepted benchmark, exceeded 10% per year in some countries such as China and the USSR, and averaged 5.5% in France. Today, growth rates vary between 0.5% and 3%[4] per year.

In other words, we've already entered a period of zero growth and in some years, such as 2020, of decline.

---

[3]. This aggregate is calculated as the sum of gross value-added (final output minus intermediate consumption), generated by production units in a given territory in a given year at current market prices.

[4]. In the United States, 5.7% in 2021, and -3.4% in 2020.

It's this tendency that leads declinists to call for more and more measures to combat the many excesses associated with economic development. But instead of doing so for moral reasons, which would be entirely justified, they invoke survival—if we don't reduce the rate at which we violate natural laws, we will die along with the world we will have destroyed.

This argument is no longer aimed at the individual but at all of humanity. But it seems to overlook the fact that there will soon be 10 billion of us on earth, and that we'll have to explain to the billions who don't have access to basic necessities that they shouldn't rebel against those who do, and who have, in a sense, denied them access. It also ignores the fact that history is littered with fears and anxieties in the face of seemingly dead-end developments, while ingenuity almost always makes it possible to cope.

The British economist Thomas Robert Malthus (1766-1834) predicted the apocalypse, or something close to it, when he showed that population grows geometrically while wealth grows arithmetically[5]. He simply failed to see one of the enduring features of history: When faced with the greatest challenges, the human spirit prides itself on rising to the occasion. Indeed, increasing productivity and improved technology have transcended mathematics and Malthusian pessimism.

Faced as we are with a great planetary challenge, we can't look to the past and accept regression and its attendant violence. The only way out of this decline is to

---

[5]. In a geometric sequence, each term allows you to deduce the next by multiplying it by a constant called "reason." In an arithmetic sequence, the deduction is made by adding this constant. The former grows much faster than the latter.

bring the environment into harmony with human activity; in other words, to make nature capable of removing waste and at the same time give a new impetus to economic activity. The next part of this book shows that we have the means to do this, starting now.

II. A COMPLETELY DIFFERENT APPROACH

4. Getting Back on Track

The previous chapter ended on a hopeful note. This one begins with a solution.

The ternary model—take, make, dispose—can't be questioned. It's the model of life. Since energy is the key element of all development, at whatever level, we must turn to it to build an exciting future.

We can no longer rely on coal, petrol or gas, which do not meet the imperative of reconciling nature and the economy, but we must move towards a new world in which non-renewable energies are condemned. This is what is known as the *energy transition*.

It's a term that has become part of everyday language, proving that social awareness— or the state of opinion at a given moment—changes profoundly over time. We all need to move away from fossil fuels. They still power cars (gasoline), ships (fuel oil, diesel, light marine diesel...), and airplanes (kerosene)—aviation accounts for 8% of final oil consumption and shipping for just under 7%, but as a whole, road transportation accounts for more than 49%. In the economic history of the 20th century, railroads were gradually electrified in many parts of the world. Why shouldn't the same be true for cars, motorcycles, trucks, coaches, buses, tankers and all types of aircraft?

Granted, wind power and photovoltaics offer solutions despite their intermittent nature, but they should be seen in the context of our needs, which are immense—the superlative is not an exaggeration.

On the one hand, the only way to reduce waste seems to be to slow down our use of nature. In short, we have to take less in order to throw less away. On the other hand, we have to find a way to respond to two trends that seem irreversible: population growth and demand for electricity, both of which are global.

Regarding the first, the most credible projections are that the population will reach 10 billion by 2050. As for the second, my personal estimate is that it will at least double in developed countries, but more likely quadruple globally over the same period.

Before we get into more dizzying figures, let's get an idea of what a kilowatt is. The watt is the unit of power. Energy is measured in kilowatt-hours, where one kilowatt is equal to 1,000 watts. One kilowatt-hour is equal to one kilowatt of power for one hour[1].

For each of us, a unit of measurement is abstract and we find it hard to imagine what these numbers represent. So let's try to be specific.

> "For all individuals that work in the energy sector, kWh (or, for those twisted Anglo-Saxons, BTUs) are as familiar as liters of water to the gardener, or kg of flour to the baker. But actually, when thinking about it, no one knows what a kWh really is through his (or her) senses.... Actually, the energy unit that each one of us knows the best is not the kWh, but, most probably… the Calorie[2]."

---

[1]. The kWh formula is: [number of hours of use] × [number of days of use] × ([device wattage]/1000) = number of kWh.

[2]. See Jean-Marc Jancovici, <https://jancovici.com/en/energy-transition/energy-and-us/how-much-of-a-slave-master-am-i/ >.

To help provide an explanation, Jean-Marc Jancovici, a French engineering consultant and energy and climate expert, converted one form of energy (calorie) into another (kWh): If a sedentary person needs 2,000 calories a day to live, this means that he or she consumes about 2.3 kWh a day at rest.

The website of the EDF (France's largest electricity supplier) offers a number of comparative analyses. One kWh is enough to:

– watch television for 3 to 5 hours
– work on a computer for half a day
– use a hairdryer for half an hour
– run a 1,000W radiator for 1 hour
– run a washing machine cycle
– charge a smartphone for about 4 months
– light up a house for 1 to 1 1/2 days
– drive a small electric car for 2 kilometers

A simple shower uses 2 kWh and a bath, 4 kWh. More generally, someone in Europe consumes more than 36,000 kWh per year, while a resident of Qatar uses 194,222 kWh, an American uses 78,754 kWh, a Chinese person uses 31,000 kWh, a Tanzanian uses 907 kWh and a Somali uses 217 kWh[3]. The world therefore faces a huge challenge over the next 25 years. If we are to combine our efforts to install renewable energy and nuclear power (18 billion megawatts per hour), minus the necessary reduction in fossil fuels (-11 billion), with global demand estimated at 105 billion in 2050 (30 billion in 2020), we will have to

---

3. See: <https://ourworldindata.org/grapher/per-capita-energy-use>. It's important to note that, according to this source, global demand has grown by more than 60% on average over the last 10 years.

find 105 - 30 - 18 + 11 = 68 billion megawatts per hour[4] as summarized in the graph below.

As an indication, the capacity of a large EDF plant (such as Flamanville in northern France) is 1,600 megawatts, which means: 1,600 × 24 (hours) × 365 (days) = 14,016,000 megawatts per hour. To provide 68 billion megawatts per hour, we would therefore need 68,000,000,000/14,016,000 = 4,851 power plants of this type to meet the world's annual needs[5].

It's easy to see why we need to mobilize all sources of decarbonized energy production as quickly as possible.

It should also be noted that increasing the number of conventional power plants, known as Generation III, would lead to increased radioactive waste, and we would be in a delicate race to the finish line, so to speak.

---

[4]. Information from the IEA (International Energy Agency), Enerdata (www.enerdate.net), RealEnergy (www.realenergy.com), Planète Energies (www.planete-energies.com).

[5]. According to the International Atomic Energy Agency, there are currently 438 power reactors in operation around the world. This number is expected to double in 25 years, as it takes 10 to 15 years to build a power plant.

4. Getting Back on Track

The exploding demand for electricity between 2020-2050

The gap between electricity supply and demand from today to 2050

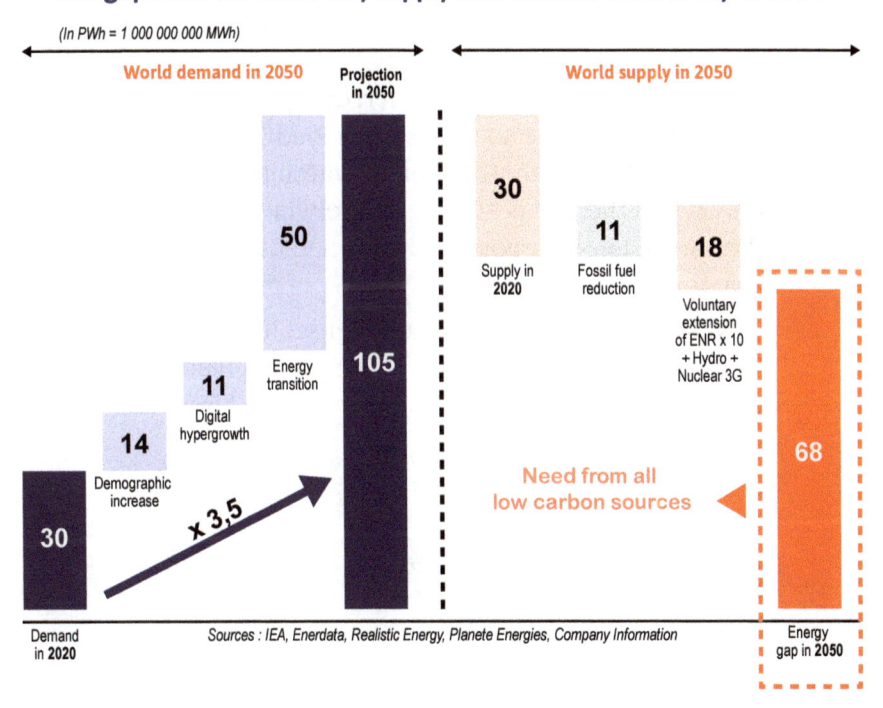

(In PWh = 1 000 000 000 MWh)

World demand in 2050 | World supply in 2050

- Demand in **2020**: 30
- Demographic increase: 14
- Digital hypergrowth: 11
- Energy transition: 50
- Projection in 2050: 105
- × 3,5

- Supply in 2020: 30
- Fossil fuel reduction: 11
- Voluntary extension of ENR x 10 + Hydro + Nuclear 3G: 18
- Energy gap in **2050**: 68

Need from all low carbon sources

Sources : IEA, Enerdata, Realistic Energy, Planete Energies, Company Information

Of course, everything would be different if it were possible to absorb very long-lived waste by recycling almost all of it. Building on the progress made in research, in particular the two prototypes of French fast-breeder reactors, Superphénix and ASTRID (Advanced Sodium Technological Reactor for Industrial Demonstration), could have shortened the time needed to achieve such a result. Between 1997 and 2000, policies and short-sighted decisions prevented us from building on this momentum towards a circular economy[6].

As a result, most industries have become increasingly dependent on electricity, which is a great thing in and of itself, provided the electricity can be produced in large quantities and is not intermittent: Who would be willing to give up the cell phone, which requires energy-intensive remote storage solutions (the cloud)? Who would be willing to give up the idea of watching television only when the sun is shining or the wind is blowing? Who would accept not knowing when they would be able to eat?

When asked about this challenge in June 2023, Xavier Piechaczyk, Chairman of the French Electricity Transmission Network (RTE), did not mince his words:

> "This trajectory raises a number of questions: will households have the appetite for faster electrification? And will the country be able to produce that much electricity? We would have to produce [...] more than twice as much as we do today. This is a real break with the past, because France today produces no more low-carbon electricity than it did twenty years ago. The availability of nuclear power stations has fallen sharply, offsetting the growth of renewables[7]."

In this context, we must be able to think outside the box that current data locks us into. Of course, we can use all the means at our disposal to increase the electrification capacity of a country like France, starting with wind turbines and photovoltaics, but nuclear power itself

---

[6]. On this subject, see the French parliamentary report no. 1028 on the causes of France's loss of energy sovereignty and independence, rapporteur Antoine Armand, March 30, 2023.

[7]. *Le Monde*, June 7, 2023.

must become clean. It can only do so by recycling the very long-lived waste that traditional power plants still produce (mainly plutonium and the minor actinides[8], which remain radioactive for hundreds of thousands of years).

Here's what we're talking about. Technologies now exist to produce *abundant, sustainable* nuclear energy. The situation has changed. It's no longer a question of facing the future with trepidation, weighed down by the fears of the old world, but facing it with determination.

---

[8]. Radioactive elements generated by the neutron-capture process but without nuclear fission, such as neptunium, americium and curium.

5. The New Order of Things

To break out of the hellish cycle that is rapidly driving us into an increasingly unlivable world, we need to produce energy with the following properties.

1. Energy must be abundant

Let's not forget: The world's population will soon reach 10 billion people, each of whom will consume on average dozens of kWh per day. So our future energy systems must be able to keep up with demand.

The more people on the planet, the more we will need to transform the current economic system, which has failed to halt the decline in living conditions around the world. Extracting water and making it drinkable, as well as producing carbon-free vehicles (cars, motorcycles, buses, trucks, ships, airplanes, etc.) requires a lot of energy. How can we ask people to switch to electric cars as soon as possible without making them affordable?

2. Energy must be accessible

Energy must be able to provide electricity or access to low-carbon heat anywhere on the planet, even in the most remote areas.

This directly affects networks and isolated areas (small islands, for example): For energy to be abundant, we need transportation not to reach a saturation point.

In Europe, we are approaching such a limit. As far as remote areas are concerned, we need energy sources that can be installed anywhere, independently of transport networks, and that can also be connected to them.

3. Energy must be affordable

This means that high costs must not lead to the exclusion of the poorest. For example, in September 2022, the cheapest offer in France was €0.20 per kWh, in Germany €0.51, in Japan €0.53 and in the United States €0.13[1].

If the price of energy gets too high, we're back to the original problem: How can we ensure survival, life and prosperity if the cost is too high? Again, the 17 Sustainable Development Goals set by the United Nations are extremely ambitious. If they only concerned the world's most advanced countries, it would be conceivable to charge higher prices for the energy needed to keep their economies running, but that would mean a reduction in living standards. But we're not talking about a few developed countries, we're talking about the whole world. Churchill was right when he said, "You don't make the poor richer by making the rich poorer[2]." If energy isn't affordable, you might as well say there isn't any!

4. Energy must be clean

This is not the case with coal, oil and gas, which have made undeniable progress but at the cost of unacceptable

---

1. <https://www.fournisseurs-electricite.com>. These are medium-term retail market prices, not the European spot market price, which has risen above 1,000/MWh, or €1 per kWh.
2. On November 15, 1919.

environmental degradation. In particular, the production of CO_2 condemns these sources at a time when decarbonization of industry is the watchword. As for today's nuclear power plants, although they don't produce carbon dioxide (and this is absolutely essential), they require large quantities of water to cool their reactors, and their use of uranium produces radioactive waste, which is only partially recycled and stored in "pools" that retain their radioactivity for hundreds of thousands of years.

5. Energy must be produced safely

The three accidents at the well-known nuclear power plants (Three Miles Island in 1979, Chernobyl in 1986 and Fukushima in 2011) have had a lasting impact on public opinion. Despite the fact that, with the exception of Chernobyl, no deaths were directly attributable to these incidents, they have undoubtedly increased the fear of nuclear power among certain segments of the population. But they have also helped to strengthen measures designed to make nuclear power plants safer. Given France's level of requirements (known as its safety profile), none of these accidents could have happened in France.

6. The energy supply must be secure

This reflects the concern of every country in the world to ensure reliable access to energy sources. From this perspective, the conflict initiated by Russia in Ukraine in 2022 has reinforced the importance of this imperative. The concepts of independence and sovereignty are inextricably linked, and the general public should no longer see its energy supply as dependent on geopolitical or macroeconomic situations that are beyond its control.

These different imperatives led the World Energy Council (WEC)[3] to focus its 2010 report on assessing national energy and climate policies from a sustainability perspective. In an analogy to the word "dilemma," the report uses the term *energy trilemma*.

This is based on three main objectives: *sustainable energy* to combat climate change; *accessible energy* for universal access to development; and *secure energy* to prevent malfunctions.

Responding to these three objectives, which in many ways can be contradictory, is a challenge for those involved in the energy "sector." It could hardly be more in line with the global ambition set out in the United Nations' Agenda 2030 to "free the human race from the tyranny of poverty and want, and to heal and secure our planet," all the more so because accessibility carries with it another, more powerful notion, that of *energy equity*. Each of us knows very well that it would be unfair not to have access to medicine, for example. To be without electricity today is to be quickly marginalized.

Sustainable, fair, safe! Does this kind of energy exist? Is it even conceivable?

Before explaining why the answer is "yes," we must state loud and clear that there will be no way out of the current impasse without meeting the six conditions listed above. If each of these conditions is in place, not only will there be a change of course, but a new era will be ushered in.

---

[3]. Founded in 1923 as the World Electricity Conference, it is closely linked to the UN's Agenda 2030 and its 17 Sustainable Development Goals (SDGs).

5. The New Order of Things

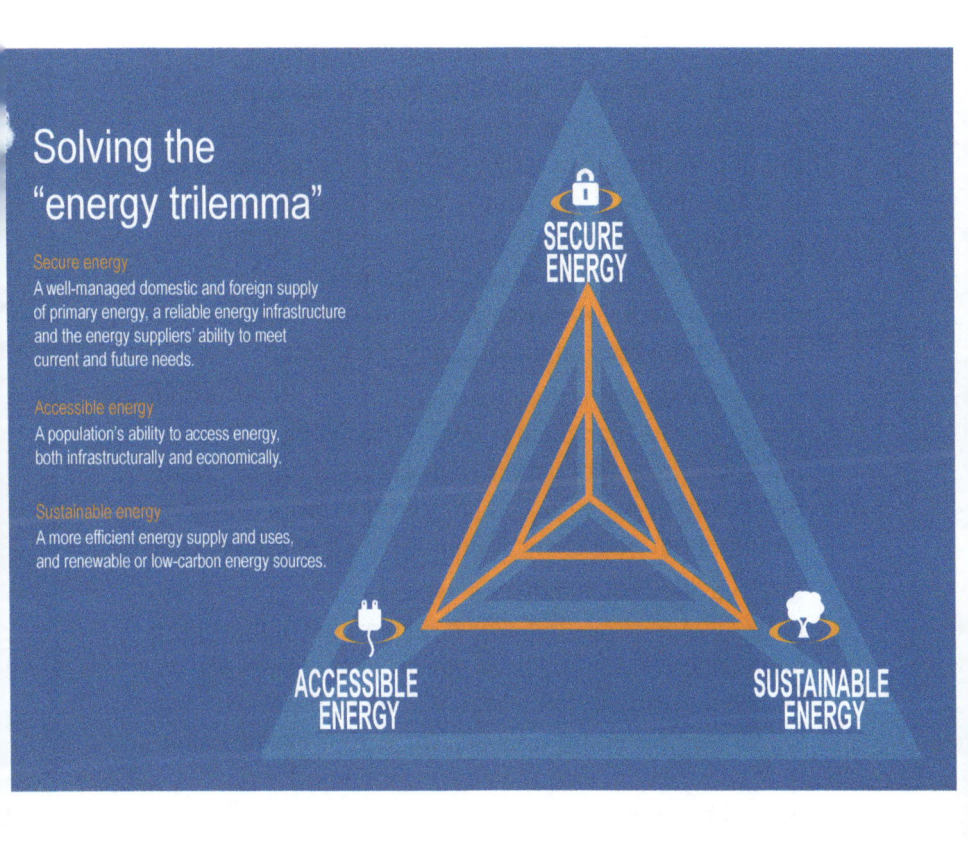

Solving the "energy trilemma"

Secure energy
A well-managed domestic and foreign supply of primary energy, a reliable energy infrastructure and the energy suppliers' ability to meet current and future needs.

Accessible energy
A population's ability to access energy, both infrastructurally and economically.

Sustainable energy
A more efficient energy supply and uses, and renewable or low-carbon energy sources.

- SECURE ENERGY
- ACCESSIBLE ENERGY
- SUSTAINABLE ENERGY

Neither coal, oil, gas, wind, photovoltaics, hydropower, biomass (any organic matter that can be converted into energy, such as wood, biogas, or biofuel) nor traditional nuclear power meets these three conditions; each can meet only one or two of them. The first three are not sustainable; the next two are not sufficiently reliable (regularity of supply); hydropower and biomass are not equitable (given their uneven geographical distribution); the last is not clean in that it produces waste that cannot be recycled for the time being.

Let's assume that an energy source is capable of almost completely recycling long-lived and highly radioactive nuclear waste; that it will be able to do so not in a few decades but in the very near future (2030); that the electricity or heat it would produce would be accessible throughout the world, including the most remote areas, of course; and that it would deliver kilowatt hours at a competitive price. In that case, such an energy should be implemented immediately.

The good news is that this energy exists, and it's within our grasp. The technology exists to harness it. It's a technologically advanced form of energy. Failure to create the conditions for its use in France and Europe would be tantamount to depriving humanity of a legitimate hope.

If we don't use it, how can we explain to present and future generations that we've deliberately ignored the most effective solution to all our problems? How can we justify not using it? What can we say to those who will tell us that we had the means to correct our mistakes, that we knew it, and that we condemned those who came after us?

Given the current situation, we cannot and must not give up. The following chapter shows how we can concretely address one of the greatest challenges we have ever faced: *save the planet* (to use a common expression), which means trying to *save humanity*.

This may sound grandiose, but being grandiose is not to be confused with pretentious. This pragmatic approach—the opposite of declinism—places action in a double perspective:

– contributing to *respect for the environment* and its restoration (elimination of long-lived radioactive waste and decarbonization of energy)

– participating in *a new economic prosperity* as opposed to degrowth (abundant energy).

6. A Pioneering Spirit

Given the conditions outlined in the previous chapter, the following guidelines must satisfy the *safety-accessibility-sustainability* trilemma:

1. Abundant, non-intermittent energy is an indisputable argument in favor of nuclear power, provided that it is sustainable, that is, that it recycles long-lived radioactive waste;

2. Accessibility, on which an equitable energy system depends, implies that power plants can be built in all areas, including the most inaccessible, at an affordable price;

3. Safety[1], which requires proven technology.

Together, these three things form a unique set of issues that need to be addressed. The previous section dealt with the issue of nuclear power generation. Those reading this book will not be offended, I hope, if I remind them of a simple but essential principle of physics.

1. In the universe, every entity (things and living matter) is made up of atoms, and every atom is made up of a nucleus around which particles, electrons, gravitate. The electrons are bound to the nucleus by a force that prevents them from escaping. Inside the nucleus

---

[1] *Safety*, which is an internal factor in any system, should not be confused with *security*, which aims to prevent and eliminate any external threat.

are other elementary particles: protons and neutrons. Protons have a positive electric charge, while neutrons are neutral. Electrons have a negative charge (electric charge is a property that leads to interactions between different bodies).

In nature, two particles with the same charge repel each other. To prevent the nucleus from exploding under the pressure of the protons, an immense force must act to hold them together (neutrons, as their name suggests, do not act).

To illustrate, the membrane of the nucleus is so strong that it will not give way unless we succeed in making it burst, which is particularly difficult but not impossible. The energy of the elementary particles (called the "binding energy") inside the nucleus is several million times greater than that which binds the electron to its nucleus. If we succeed in breaking this "packaging," the nucleus bursts, releasing neutrons and considerable energy in the process.

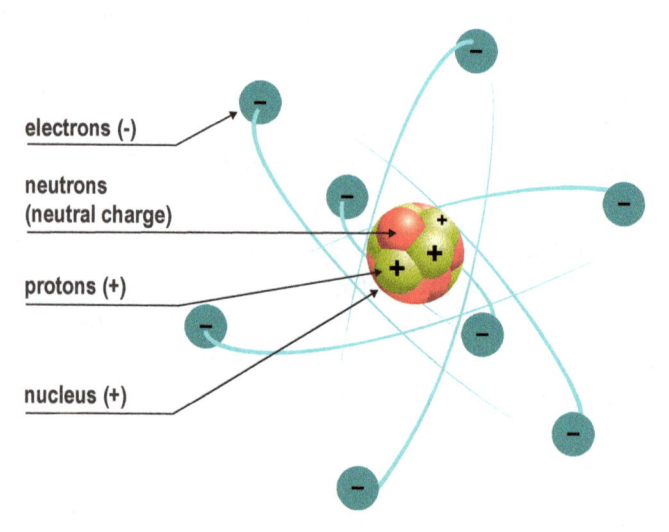

electrons (-)

neutrons
(neutral charge)

protons (+)

nucleus (+)

6. A Pioneering Spirit

Since Henri Becquerel's discovery in 1896 that uranium emits radiation on its own, and since Marie Curie's work on radioactivity, it is this force that the whole world dreams of controlling. The way matter is organized means that the splitting of an atomic nucleus releases more energy than a simple chemical reaction when it burns. In short, a very small amount of uranium produces more energy than a large amount of coal: 1 gram of uranium is equivalent to 3 tons of coal (3,000,000 grams), or a ratio of 1:3 million (the ratio for gasoline is 1:2 million).

When the atom splits, it releases the energy it contained up to that point, commonly known as "nuclear energy" or "atomic energy." The problem then becomes how to control the reactions caused by this explosion, because nucleons (neutrons and protons) launched at very high speeds are not easy to control.

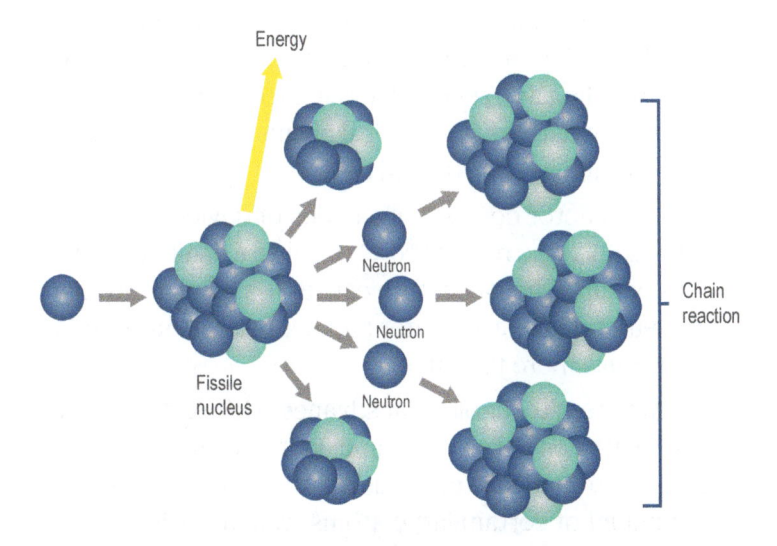

Nuclear fission

After this process was discovered, science first used it for military purposes. The two atomic bombs dropped on Hiroshima (August 6, 1945) and Nagasaki (August 9, 1945) are still fresh in everyone's memory. But the world also discovered a wide range of civilian uses for nuclear energy.

Today, much like cell phones (from the first generation to 5G), a new generation of nuclear reactors is emerging, the so-called fourth-generation *fast neutron* reactor.

All techniques for harnessing nuclear energy must take into account one natural fact: radiation. From natural uranium, the most abundant in our country, we can extract uranium-235, which is fissile (it can sustain a nuclear chain reaction). But 100 grams of the former contain only 0.7 grams of the latter (0.7%). Uranium-235 must then be concentrated into natural uranium to reach 4%, which is the critical mass used in a pressurized water reactor. The first fission of concentrated uranium-235 is achieved by neutron bombardment. The neutrons are propelled at a speed of about 20,000 kilometers per second. To harness this physical phenomenon, it is necessary to build a sophisticated device: the nuclear reactor.

The reactor consists of a series of devices that use nuclear fuel (the reactor core) to cause a chain reaction. This occurs when a neutron causes fission of an atomic nucleus, releasing a greater number of neutrons that in turn cause more fission.

The great technological advance of the 20th century was man's ability to initiate and control chain reactions, either automatically or on his own. In a nuclear reactor, the nuclei of certain large atoms (called fissile atoms) can be split in two by a collision with a neutron. When the neutron strikes the nucleus, the heat released can be extracted and converted into electrical energy. The primary fuel used is uranium-235, the only fissile isotope of natural uranium[2].

The first generations of nuclear reactors were designed to slow down neutrons considerably by placing an obstacle in their path. Some materials are good "speed moderators," such as hydrogen or deuterium (one of its isotopes), or graphite. Graphite slows down neutrons without trapping them. The first power plants (and the current third generation) use slow neutrons, no longer at 20,000 km/s, but at about 2 or 3 km/s.

In France, for example, the graphite gas process was first introduced using pressurized carbon dioxide to cool the circuit. It was abandoned in 1969, followed by a second generation using pressurized water for heat removal, and then a third, also using pressurized water, called the EPR (European Pressurized Reactor, later renamed Evolutionary Power Reactor), which is more powerful and resulted from a Franco-German partnership. The fourth generation is now working on the development of *fast neutron* reactors[3].

2. *Accessibility* can only be achieved if the natural conditions of geography do not stand in the way of equality. For example, some parts of the world have no water.

---

[2]. Natural uranium consists of three isotopes (elements with the same number of protons in the nucleus): uranium-238, the heaviest and most abundant; uranium-235, and uranium-234. Uranium-235 is the only isotope of natural uranium that undergoes fission, so it is widely used. Uranium was named after the planet Uranus, which was discovered in 1781.

[3]. There are currently six Generation IV reactor technologies, five of which use fast neutrons. They differ in two main ways, the type of core cooling and the type of fission. The only technology that uses liquid nuclear fuel is molten salt, which acts as both a heat transfer medium and as a containment barrier.

Others are islands or areas where it's not possible to install electricity transmission networks. In all these cases, proximity to production is a key advantage.

In a collection of essays published in 1973[4] under the title *Small is Beautiful*, the British economist Ernst Friedrich Schumacher defended the idea of an economy on a human scale. At the time, for example, computers were still huge, and the prospect of tablets or mobile phones was almost unimaginable.

So far, gigantism has prevailed in the field of energy production. Not only in terms of size (two to three hundred hectares for an EDF power plant, for example), but also in terms of power generated (from 850 to 1,600 megawatts). Today, a complementary approach that takes advantage of small size is gradually gaining ground. Many countries (USA, Canada, Russia, China, Korea, India, France) have joined the race with increasingly convincing results. Recent years have also seen the development of what experts call SMRs (small modular reactors) and AMRs (advanced modular reactors). Both are small reactors with an electrical output of between 50 and 350 megawatts[5].

The trend is gaining momentum in France, in particular with companies such as Naarea[6]. Naarea has developed a fourth-generation microreactor, the XAMR[7] (eXtrasmall Advanced Modular Reactor), which is ultra-compact and has an electrical capacity of 40 megawatts (therefore

---

[4] *Small is Beautiful: A Study of Economics as if People Mattered*, 1973.

[5] 1 megawatt = 1,000 kilowatts, or 1 million watts. 1 terawatt (1 TW) = 100,000,000,000 W, or 100 billion watts.

[6] Naarea, which stands for Nuclear Abundant Affordable Resourceful Energy for All, was founded in February 2020.

[7] X for eXtra small.

lower in power than the smallest modular reactors currently in use anywhere in the world). It is a fast neutron system using molten salt technology.

In the XAMR microreactor, radioactive fuel is bombarded with neutrons. This results in a self-controlled reaction: Below a certain equilibrium temperature, the cooler salt becomes colder, contracts and increases the density of radioactive nuclei, which in turn increases reactivity, the number of fission reactions and therefore the temperature; above the equilibrium temperature, the warmer salt expands and the density of radioactive nuclei decreases, with the immediate effect of reducing reactivity and the number of fission reactions, causing the temperature to drop.

The result is that below a certain temperature nothing happens; above it everything stops. This automatism is a factor that considerably increases the safety of the system, which naturally maintains equilibrium.

The inherent dynamics of molten salts are therefore much more favorable from a safety point of view than those of a solid fuel reactor. To illustrate the difference, imagine this. With molten salts, everything happens as if a marble could move up and down in a cup, depending on the temperature, and always return to its original position. With solid fuel, imagine a small hill with the same marble on top, which must be prevented from rolling to one side or the other. The second exercise is much more difficult than the first, which occurs automatically.

In addition, there are three other elements that increase safety while increasing energy equity.

The XAMR *uses no water,* so there is no risk of explosion.

Because the pressure in the structure is close to that of the atmosphere, the residual heat is simply cooled by air.

The liquid state of the fuel allows a quick response to power demand (it is capable of increasing or decreasing power in a fraction of a second) and it can be emptied in the event of an emergency shutdown.

Together, these features give this technological solution a remarkable passive safety profile[8].

In short, the XAMR combines three major innovations that have been the subject of extensive research in the nuclear industry: molten salts, fast neutrons and modular reactor design.

These considerations are somewhat technical, but it's important to understand the nature of this "new" technology. In fact, it's only really new in terms of how it is used, as it dates back several decades. The concept of molten salt reactors dates back to the 1950s, particularly with the work of American physicist Alvin Weinberg. The first prototype was built and put into operation in 1954 in Oak Ridge, Tennessee. A second reactor of this type, the Molten Salt Reactor Experiment (MSRE), operated between 1965 and 1969 and demonstrated the potential benefits of this type of device, in particular its intrinsic safety.

Fast-neutron reactors first appeared in the 1940s. In 1951, the first experimental breeder reactor (EBR-1) was also built in the United States, in Idaho. It produced electricity for non-commercial purposes for more than 30 years.

---

[8]. *Passive safety* refers to all measures in a product which, by their presence or when triggered, prevent the risk of malfunction and minimize the severity of an accident.

The abandonment of these techniques was due to one important factor: the need at the height of the Cold War to obtain a nuclear weapon as quickly as possible and to produce large quantities of plutonium to fuel the nuclear deterrent. In a pressurized water reactor using slow neutrons, the doubling time for this element is much faster than in any other reactor. For budgetary reasons, the choice of this last technology was imposed and immediately buried all other initiatives.

France has historically been one of the pioneers in the field of fast neutron reactors, with the Phénix (built in 1973 and operated until 2010 at the Marcoule site in the Gard department), and then the Superphénix (at the Creys-Malville site in the Isère department). Since the beginning of the 2010s, there has been renewed interest in the Small Modular Reactor (SMR) concept, because of its potential to reduce costs and timescales, as well as its flexibility in adapting to different energy needs.

In addition to the safety of this small device, it has the added advantage of being suitable for industrial production. Mass production reduces design and manufacturing costs per unit as well as assembly time, thanks to standardization. Because it is already intrinsically safe, its equipment and infrastructure costs will be lower, making electricity and heat more affordable.

3. Finally, the XAMR does not produce highly radioactive long-lived waste, opening the way to *sustainable nuclear energy*.

In fact, the great advantage of combining fast neutrons with molten salts is the type of fuel that can be used—plutonium and actinides—especially since there's no need to slow down the active neutron(s). It then becomes possible to use spent nuclear fuel, running on radioactive waste from the bombardment of uranium. In other words, this technology makes it possible to significantly reduce and even eliminate very long-lived, high-level nuclear waste by closing the fuel cycle, that is, by reusing all waste until it disappears.

Devices such as the XAMR open up new prospects for the recycling of radioactive waste. The spent fuel stored in France, for example, will provide a reserve for at least several hundred years. Without reprocessing, it will have to be stored underground until it is neutralized by time (more than 100,000 years). It is now possible *to close the fuel cycle completely* (99.99%) and *promote the circular economy*. In other words, eliminate waste.

> "However, since Enrico Fermi, all physicists know that the future lies with fast neutron reactors, in which neutron slowing is avoided. These reactors are capable of converting 99.3% of U 238 (fertile) into plutonium 239 (fissile). The Pu 239, previously considered as a troublesome waste product, becomes a useful fuel. The amount of energy that can be extracted from natural uranium is thus multiplied by a factor of 50 or more. Under these conditions, the specter of a nuclear fuel shortage disappears[9]."

This is a key point. The recycling of waste, which turns it into a resource, not only has the same significant advantage of switching to renewable energy, but also provides reserves that ensure national independence and sovereignty for a long time to come.

---

9. Jean-Claude Rousseau, engineer at the French Atomic Energy and Alternative Energies Commission (CEA), "Nuclear Energy: A Brief History," December 14, 2015. See: https://www.encyclopedie-energie.org/en/nuclear-energy-brief-history/.

century: reconciling energy abundance (an equitable energy system), security (resilience and sovereignty) and environmental sustainability (use of nuclear waste as fuel and decarbonization).

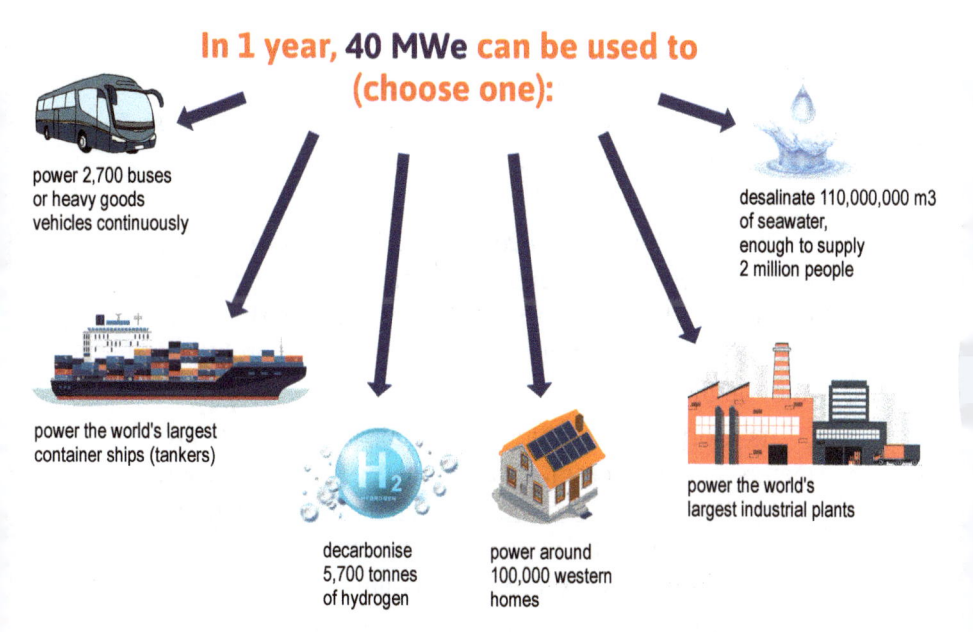

In 1 year, 40 MWe can be used to (choose one):

- power 2,700 buses or heavy goods vehicles continuously
- desalinate 110,000,000 m3 of seawater, enough to supply 2 million people
- power the world's largest container ships (tankers)
- decarbonise 5,700 tonnes of hydrogen
- power around 100,000 western homes
- power the world's largest industrial plants

Depending on how quickly the French Nuclear Safety Authority (ASN) grants its approval (generally three to five years), industrial-scale production could begin by 2030. In France, the XAMR is already in a class of its own and foreshadows a likely development on a global scale.

In short, the XAMR eliminates the need to extract materials from the ground or subsoil and enables the absorption of existing waste. In other words, it radically alters the *take-make-dispose* triptych by eliminating the first phase and reducing the third phase.

In essence, this type of very small reactor is designed to be modular; it is factory-built, can be assembled on site and is easy to install; and it can be installed close to users in locations with different configurations to best meet local energy needs.

It opens the door to *decentralized energy production,* which strengthens sovereignty through autonomy. Since it does not depend on the distribution grid (although it can be fed into it), it eliminates transmission problems and significantly increases user resilience[10].

Its footprint is particularly small, about 20 square meters for the reactor itself and about 50 for its container, which is equivalent to the size of a modest apartment and can be buried if necessary. The impact on nearby living ecosystems is therefore very limited.

This sustainable, abundant and autonomous energy allows a territory or a country to gain sovereignty[11]. The choice of molten salt and fast neutrons, combined with the small size of the reactor, therefore satisfies all three conditions of the energy trilemma.

As a result, the 40-megawatt XAMR is a tool that will enable us to meet one of the major challenges of the 21st

---

[10]. From a psychological point of view, this word expresses the ability to overcome traumatic shocks. In physics, it refers to the resistance of a metal.

[11]. In its report of July 8, 2021, the French Senate recognized all the advantages outlined above. See "Nuclear power of the future and the consequences of abandoning the 'Astrid' project," by Thomas Gassiloud, deputy, and Stéphane Piednoir, senator.

III. RETHINKING THE FUTURE

7. The International Competition

Faced with the climate emergency, more and more private actors are joining forces with hope and confidence to lead humanity into a new era: that of *massive "energization."*

To meet the ever-increasing demand for electricity and decarbonized heat, we need to completely rethink our supply policies: production can no longer be based on fossil fuels, as was still the case in 2023 for 61% of the world's electricity[1].

The transition to totally clean energy is not just a possibility or a wish but, to borrow a phrase from De Gaulle on planification, an 'ardent obligation.' Sustainable nuclear energy is the answer. It's the only way to accelerate the transition to all-electricity, even though other decarbonized technologies can also contribute.

Specialists and ordinary mortals alike know that electricity is magic, as we might have understood from the French painter Raoul Dufy[2]. Not because it comes from

---

1. According to a report by the Ember think tank in the *Global Electricity Review*, April 2023, which analyzed data from 78 countries representing 93% of global electricity demand.
2. Dufy was commissioned to create a huge mural for the Pavilion of Light and Electricity at the 1937 International Exposition in Paris, which he titled *La Fée Electricité*—in English, "The Electricity Fairy." The mural, which traces the history of electricity, is now in the collection of the Musée d'Art Moderne de la Ville de Paris.

some kind of wizardry, of course, but because it can change the very structure of our industrial organization.

From trains to automobiles, from airplanes to ships, from the simplest household appliances to the most complex electronic devices, electricity provides the most valuable services without harming nature—at least as long as its production does not generate non-recyclable waste.

An electric world would be a great thing if it were based on sustainable energy. This idea will one day prevail, but we still have a long way to go; this will require profound changes.

Internationally, at the time of writing, there is a great deal of competition between operators using molten salt, fast neutron and small modular reactor technologies. As for the XAMR, a French patent, it aims to produce heat and electricity by 2030 with a 40-megawatt reactor (80 megawatts thermal) at a lower price than any other source (coal, oil, gas, hydrogen, wind or photovoltaic). In addition to the benefits of decarbonization, this is good news for purchasing power.

The French are famous for their inventiveness and capacity for innovation, but also for their relative inability to systematically exploit their own creativity; as the British writer Rudyard Kipling quipped, the French are "first to face the Truth and last to leave old Truths behind." He wasn't wrong, if we take a quick inventory of the French inventions that others, starting with the Americans, have turned into supply shocks; in other words, major impacts on all sectors of the economy[3].

---

[3]. First piston steam engine, Denis Papin, 1687; first motorized land vehicle, Joseph Cugnot, 1770; first photograph, Joseph Nicéphore Niepce,

7. The International Competition

All French inventors could have gotten a head start—if they had known how to systematically transform these innovations into industrial assets.

In the nuclear field, the country has sabotaged itself. The Superphénix project was the prototype of a fast breeder reactor (RNR). With a capacity of 1,240 MW, it was built in 1976, activated in 1984 and shut down in 1998. The ASTRID project, launched in early 2010 and shut down in 2019, was also a prototype of a fourth-generation, sodium-cooled fast neutron reactor. A number of factors explain the cancellation of these two projects, including a lack of long-term vision, a lack of public consultation, a failure to heed the warnings of researchers and political hesitation, as highlighted by the Armand report[4]. These past experiences have made it possible to accumulate knowledge; it is highly regrettable that, despite some well-founded criticism[5], the interruption of research and experimentation has led to a loss of expertise.

Today, the country's nuclear know-how and advances in molten salts, fast neutrons and very small reactors fueled by nuclear waste can give France a double advantage: being competitive in terms of economics, and ethical in relation to the environment.

1825; first photovoltaic cell, Edmond Becquerel, 1839; first rechargeable battery Gaston Planté, 1859; first electricity transmission over a long distance, Marcel Deprez, 1881; first aircraft, Clément Ader, 1890; first movie projector, Auguste and Louis Lumière, 1894; first submarine, Maxime Laubeuf, 1900; first helicopter, Paul Cornu, 1907; first multistage rocket, Louis Damblanc, 1936; first microcomputer, François Gernelle, 1972; first smart card, Roland Moreno, 1974; the Minitel (forerunner of the Internet), 1980...

4. A 2023 parliamentary report on an inquiry led by deputy Raphaël Schellenberger with lead 'rapporteur' Antoine Armand explored the reasons for the loss of France's energy sovereignty and independence.

5. Sodium cooling is dangerous because of its flammability.

8. A Systemic Evolution

All the major power plants in the world are designed to supply electricity to large areas through transmission and distribution networks for a commodity that, it should be remembered, is virtually impossible to store on a large scale. The size of these production units (hundreds of hectares for capacities ranging from 350 to more than 1,500 MW) means that supply is concentrated in a small number of sites, as is the case in France (18 nuclear power plants with 56 reactors in operation as of January 1, 2023) and is dependent on the grid. This requires heavy and costly infrastructure, a major handicap for developing countries. The most developed countries are not exempt from the need to maintain these networks and, more importantly, to adapt them to changing demand. Given the energy gap mentioned above[1], there is a risk of saturation.

In almost all existing systems around the world, the total energy supply is made up of various centralized sources (hydro, conventional nuclear, coal, gas), and renewable and often decentralized sources such as wind power and photovoltaics[2].

Unlike wind and solar, the latter two are intermittent: They vary according to wind and solar conditions. What's more, they require heavy installation (especially

---
1. See diagram on p. 41.
2. Biomass is not included here.

offshore wind) at high cost and, in the case of solar, a large land footprint with highly variable results. Nevertheless, what we call the "energy mix" is an imperative given the gigantic needs mentioned above[3].

This global production of energy only makes sense if it can be transported to the users, whoever they may be. This is the purpose of the electricity transmission grid, which is fed by the various intermittent and non-intermittent producers. Saturation is always possible as a result of the increasing number of connections, for whatever reason: natural demographic changes, migratory flows, industrial growth, attractiveness of the country to foreign investors, growing needs, energy transition and so on.

What's more, it's difficult to know exactly what the state of the network is. Finally, the distribution network has to deliver electricity to various customers (factories and households). An increase in demand requires an increase in supply, but this immediately puts additional pressure on the transmission system, with repercussions on the distribution system, in other words, possible power cuts for everyone[4].

---

[3]. Wind turbines and photovoltaics contribute to power, although still at a low level. France produces 92% decarbonized electricity, mainly from nuclear power (71%), hydroelectric power (11%) and, to a lesser extent, wind power (6%) and photovoltaics (2%). Joint opinion of the French Académie des sciences, Académie des beaux-arts and Académie des sciences morales et politiques on February 24, 2022.

[4]. This is typical of France, where the reduction of nuclear power generation has been a goal since the 2010s. With the energy crisis triggered by the war in Ukraine, the situation has been reversed. We now need to restart closed plants and build new ones, which is never immediate (10 to 15 years).

In this context, the arrival of a non-intermittent and decentralized local supply (such as that which can be provided by an XAMR) would help to stabilize the overall balance of the system, however it is organized[5], in three complementary ways.

1. First, the system can be implemented as close as possible to the users, mainly companies and municipalities, whoever they may be. It guarantees continuity of supply, independent of the transmission and distribution network. In other words, the consumers would acquire an autonomy that guarantees their supply.

2. Then, in the event of an increase in demand (regular due to typical economic activity, or exceptional due to climatic or other conditions), this contribution would allow us to relieve pressure on the electricity transmission network and the centralized producers. It would be therefore not only *additional* but truly *complementary*, since it is compatible with the development of conventional renewable energies and nuclear power. More generally, the same effect would be produced on the interconnected networks of the European Union.

3. Finally, we must never forget that any electricity production unit, whether nuclear or thermal, produces heat. The XAMR, with a capacity of 80 thermal megawatts, reaches 650°C. This carbon-free heat can either be supplied in its entirety to a company for industrial

---

[5]. The French nuclear ecosystem, for example, is made up of six entities each with a specific role: the French Alternative Energies and Atomic Energy Commission (CEA), a public research and development organization; Électricité de France (EDF), the electricity producer; Framatome, EDF's prime contractor; Orano, which reprocesses spent fuel; the Electricity Transmission Network (RTE), France's transmission system operator; and ENEDIS, France's main distribution system operator. There are also 80 private energy companies whose main customers are businesses.

processes, or it can be converted into electricity. In the latter case, the reactor's maximum output is 40 megawatts, but its operation generates what specialists call "waste heat," which is unavoidable, because it is inherent to the energy conversion process. This waste heat, at a temperature of around 75°C, can itself be used commercially, for example to heat factories. In France, most electricity is decarbonized because it comes from nuclear power; this is not true of heat production, as is the case in most countries in the world. This comes mainly from the use of gas, not only for businesses but also for domestic use, as anyone with a non-electric water heater knows. A microgenerator of the XAMR type would therefore also supply decarbonized industrial heat (always decentralized) corresponding, for example, to the needs of light industry (such as assembly, paper/cardboard, chemicals, recycling, food processing). In this way, a strong contribution to the decarbonization of industry could be made.

So if we think in terms of the *electricity system*, the arrival of this decentralized offer, both for electricity and heat, can complete an ensemble that is already engaged in the energy transition.

The diagram below shows the result of the introduction of a new distributed generation (the XAMR) into the system.

8. A Systemic Evolution

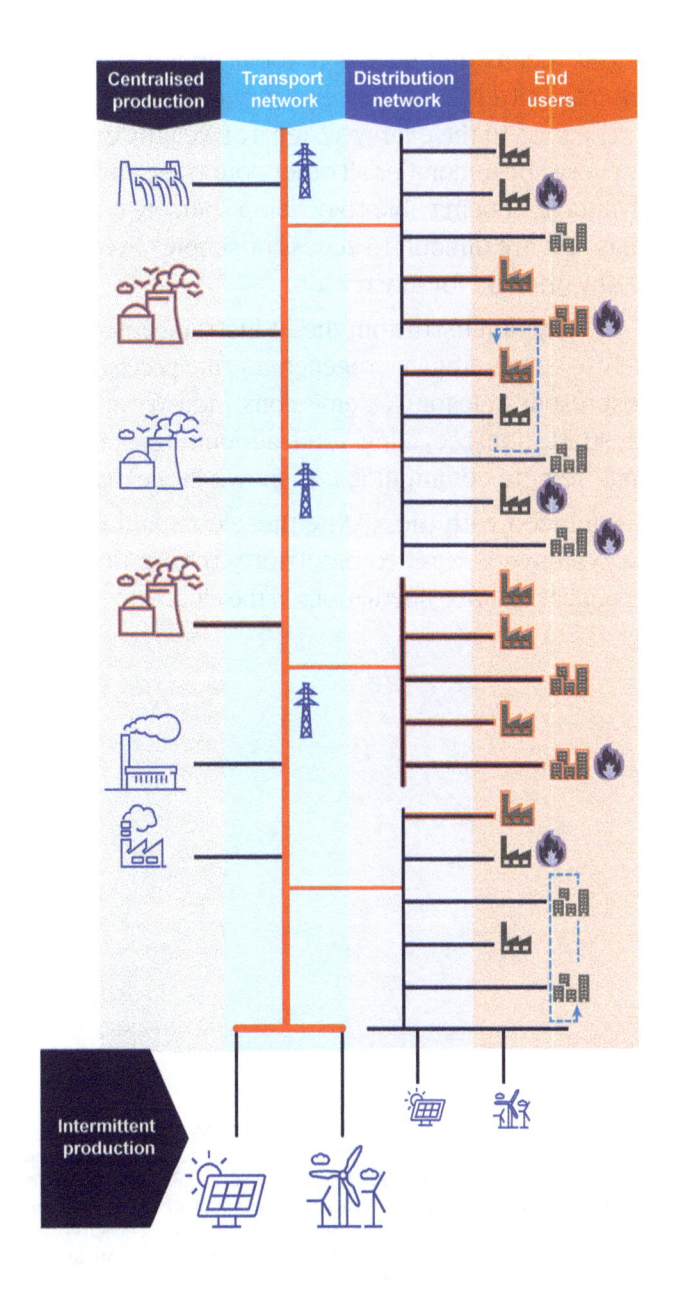

Centralised production	Transport network	Distribution network	End users

Intermittent production

This means that access to local, abundant, secure, non-intermittent, low-cost energy would make a major contribution to the energy systems of wealthy countries, for every consumer. For all other countries, it offers the advantage of being able to develop sparsely populated areas that are difficult to access or simply have no grid, for any grid and for any reason.

By way of illustration, the industrial sectors most likely to benefit from microgenerators are precisely those most responsible for CO_2 emissions: industry in general, transportation, recycling, food, agriculture, smart buildings, remote communities, as shown in the diagram[6].

Equipped with the XAMR, these companies would always have access to electricity that is no longer dependent on price fluctuations in the electricity market[7].

---

[6]. <https://ourworldindata.org/emissions-by-sector>

[7]. Surprising as it may seem, there is a market for electricity, just like any other good. Several types of transactions take place: daily, time-dependent (from one week to 3 years), bilateral (over-the-counter) and strictly regulated (ARENH - Regulated Access to Incumbent Nuclear Electricity). As an indication, the price per kilowatt-hour exceeded 1,000 euros in August 2022, while its usual level was around 40 euros.

8. A Systemic Evolution

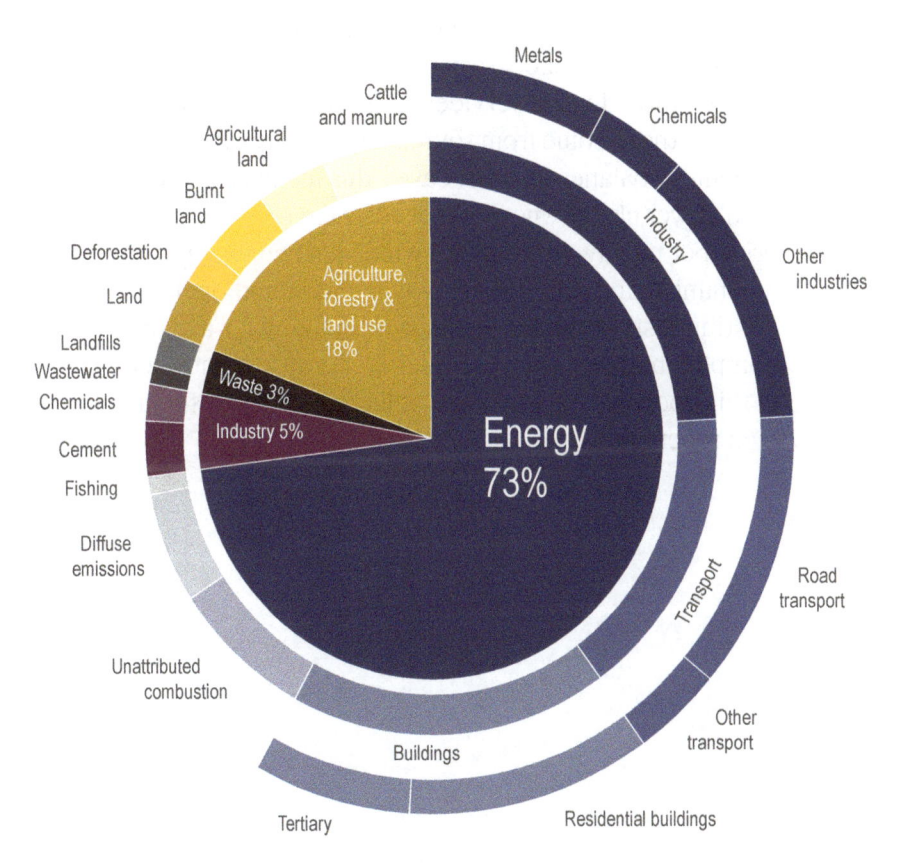

[Figure: Nested ring chart showing global greenhouse gas emissions by sector. Inner ring: Energy 73%, Agriculture, forestry & land use 18%, Industry 5%, Waste 3%. Outer ring labels: Metals, Cattle and manure, Agricultural land, Burnt land, Deforestation, Land, Landfills, Wastewater, Chemicals, Cement, Fishing, Diffuse emissions, Unattributed combustion, Buildings, Tertiary, Residential buildings, Other transport, Road transport, Other industries, Chemicals, Industry, Transport.]

In the spot power market (often called the day ahead market), trading takes place in a closed market for the following day. There's a simple reason for this—you don't have to place an order before you turn on the power. You can't say, "I need to turn on my TV (or my smartphone, or my automatic production line), please produce the kilowatts or megawatts I need." A push of a button has to be enough to turn it on. That's why it's important to always be a day ahead, so that when the time comes, at all hours of the day, the response is immediate. This is indeed the case for most of the needs to be met. When adjustments are necessary to avoid power cuts, they involve introducing services with new capacities. In general, these come from sources that can react quickly to achieve balance. In most cases, this means using gas-fired power plants, whose higher costs are reflected in the price per megawatt hour and which also emit CO_2. The mechanism for balancing supply and demand operates continuously at the European level, making for a highly complex market. With the microgenerator mentioned in this book, it's easy to make adjustments thanks to its high reactivity.

9. A Changing World

We're entering a new era of sustainable energy. I can't emphasize enough that this is not just a technical issue, but a change that will improve our daily lives.

All the polls show that two-thirds to three-quarters of the French people are still in favor of low-carbon energy production. This opinion is based on traditional reactors, so these figures, although already very high, would no doubt increase if very long-lived radioactive waste were eliminated and converted into a fuel source for the XAMR.

In a context where the production of energy from uranium has historically been associated with monumental structures such as nuclear power plants, the emergence of new energy technologies challenges this traditional approach. In contrast to the use of a centralized energy distribution system through a network, the proposed model advocates for decentralization based on proximity. This entails the ability to locate energy production as close as possible to end users. This does not challenge the concept of centralized production but offers a complementary approach. This principle of decentralization is a crucial aspect of this model.

The issue is not one of proliferation or diffusion, but rather the installation of a technology that cannot be used for military purposes and is not subject to the risk of explosion.

In order to be as close to the end user as possible, the XAMR adheres to a number of constraints[1]. It has to:

– avoid water cooling, since it must be able to be installed anywhere, even in the desert

– eliminate all harmful emissions to ensure that installation in any location will not harm the local environment

– allow a small physical footprint (less land use), since the proliferation of sites is incompatible with large areas

– adapt to the local situation of a company and/or community, whose heat and electricity needs can vary considerably

– meet the highest safety standards

– offer the prospect of accessing heat and/or electricity anywhere in the world.

Taken together, these points argue for the mass diffusion of this innovation in order to ensure that all populations have access to energy that is safe, decarbonized, non-intermittent, decentralized, affordable, clean and inexpensive. This is particularly important given that the price per kilowatt-hour is the key criterion in a rapidly developing and industrializing world.

This development does not in any way condemn existing large power plants, regardless of their location, as the use of spent nuclear fuel will gradually transform conventional nuclear power into a sustainable energy source.

---

[1]. This clarification is important because small reactors are used in nuclear submarines, for example, but they are still third-generation, meaning they do not complete the nuclear cycle and therefore produce waste that has to be stored.

France is fortunate to have a head start in this field, despite the slowdown that occurred at the beginning of the 2010s. In analyzing the reasons for France's loss of leadership in nuclear energy, the third chapter of the aforementioned report[2] is particularly relevant. It details the fact that France took approximately thirty years, from the late 1990s to the 2020s, to realize that nuclear energy could be a great asset in the face of climate change. It took thirty years to reverse the trend, despite the warnings of many specialists[3]. It took thirty years to recognize that the loss of independence was a significant threat to the country, and to understand that reducing the nuclear program would increase the risk of power cuts. It took thirty years to react!

France may have made an error in judgment with the ASTRID prototype, but since the Superphénix, it appears that the relevant lessons have been learned. The resurgence of various forms of nuclear power in the early 2020s is indicative of a shift in attitude. In Europe, France's potential to lead a divided continent remains significant[4], not only by invoking its technical advantages, as undeniable as they are (and as I believe this essay has shown), but also because the reconciliation of nuclear power and the environment represents a pivotal moment in our history. It invites us to look at the future in new ways, to think in new ways, and to recognize the potential

---

[2]. In March 2023, the French Parliament published a detailed report on the reasons for France's loss of sovereignty and energy independence.

[3]. In particular, Yves Bréchet, professor at the Grenoble Institute of Technology.

[4]. Favorable: Bulgaria, Croatia, Czech Republic, Finland, Hungary, Netherlands, Poland, Romania, Slovakia, Slovenia, France. Opposed: Germany, Italy, Lithuania, Belgium, Portugal, Denmark, Austria. About a quarter of the EU's energy comes from nuclear power, more than half of it produced in France.

of general electrification, which is not limited just to the devices and appliances we already know, from televisions and microwaves to vacuum cleaners, elevators, lighting, street signs, hair dryers, rechargeable phones, computers, tablets, tools and all the kitchen appliances we use on a daily basis, not to mention heating and air conditioning.

The world of the future will also witness the widespread electrification of a multitude of transportation modes, including trains, cars, motorcycles, buses, trucks, ships, airplanes and more. This is already happening in many places. The world of the future will be one of recycling, a circular economy, robust growth and respect for nature. A world that will bring greater freedom and simplicity to all our daily actions and to life in general. A more just world.

The ongoing fight against climate change on the one hand, combined with ensuring security of supply and therefore sovereignty on the other, necessitates a profound change in the attitudes and public policies of the most reticent countries. They must recognize the contribution of sustainable energies.

For the hundreds of millions of people who lack the basic necessities, including the number of daily calories needed to live a decent life, a 40-megawatt microreactor, expandable as needed, represents a radical change. It can provide access to potable water, which is abundant underground but must be extracted[5], purified and desalinated. It can contribute to the development of agriculture, protection against severe weather, greater mobility and communication—in short, everything that contributes to the well-being of the world's least developed countries.

---

5. Without touching the water table.

Stated simply, it would be a mistake to believe that the energy transition alone can solve the problems associated with the excesses of growth. The transition to a sustainable nuclear age can't be waved like a magic wand to solve all the problems associated with the proliferation of microparticles harmful to flora and fauna, the accumulation of plastics on a global scale, the decline in biodiversity or the extinction of certain species. A sober, more prudent form of capitalism, more cautious and aware of ecological balances, would undoubtedly be a move in the right direction. However, the transition must begin with CO_2-free energy.

As seen throughout this book, despite the implementation of various strategies aimed at reducing waste and economic greed, energy demand is projected to grow at an exponential rate in the coming decades. Without decarbonization—or low-carbon technologies such as the new sustainable energy and other renewables—we will be heading for disaster.

But as I've tried to show, the advent of this new energy provides us with a powerful tool with which to begin the process of turning hope into reality. This approach reconciles respect for the environment with the pursuit of economic prosperity. Now it's up to people and governments to accelerate the process.

Why I wrote this book

Each of us has certain tendencies, certain drives, that we're not immediately aware of, and that manifest themselves over time without our knowing how. In my case, what drove me to search for a sustainable nuclear energy gradually revealed itself until it became a kind of intrinsic motivation.

When I was young, I didn't accept that reality could get in the way of my dreams. Whatever seemed impossible was a challenge that I wanted to tackle. I was convinced that there wasn't anything I couldn't do. Of course, this was exaggerated and perhaps absurd, but what stuck with me was that before giving up, you have to be relentless. As I grew older, I realized that it takes a mixture of audacity and naïveté to set out to overcome what seems impossible. Seneca reinforced this attitude, writing, "It is not because things are difficult that we do not dare, it is because we do not dare that they are difficult."

And when I hear people say, "We've always done it this way," a little voice inside me says, "What if we did it better from now on? Why settle for what we have?" That state of mind has never left me. I don't mean to brag; it's just the way I am. This attitude led me to reject the impasse we are now faced with: degrowth or ecological disaster.

I've been struggling with this problem for a long time. I've never ceased to be an environmentalist, but I'm not satisfied with the solutions proposed by a Manichean ecology. We can no longer think of nature as a mere framework; it is an integral part of our existence. It must be restored to what it was before we began to neglect it—our habitat, a vital resource, something that underpins the very fabric of our lives. In short, we must stop thinking of the environment as an externality and start thinking of it as an internality. In this way, we can innovate to respect natural balances while improving our living conditions.

Throughout my career, I've worked in many different countries and cultures, and I've seen poverty, hunger, deadly disease and human despair that can only be remedied by improving people's living conditions. In Malaysia and Singapore, I also saw the devastation that can result from building a high-speed rail line and electrifying it through a two-million-year-old forest. That was in 1996. I saw the same kind of damage in Brazil between 1997 and 2001. I understood then how deforestation linked to the arrival of electricity in certain areas can harm fauna, flora, the atmosphere and all living things.

Despite everything, I've always believed that the human spirit is able to face the most difficult challenges and come out on top. Sometimes it has taken time, and there have been frustrations, discouragements and setbacks, but it's always been possible to push back the boundaries of the impossible. Being optimistic always seemed to me to be something to strive for.

But not blind optimism—the kind of optimism that accompanies action and is often born of it. I wrote this book because I couldn't accept the prejudices of the anti-nuclear movement, which were at odds with a legitimate desire to save the planet. Nor could I accept the prevailing pessimism, fueled by a lack of action that I found reprehensible, despite the existence of solutions that would allow us to avoid ecological disaster. I wrote this book in the hope that each and every one of us would share my refusal to accept any kind of fatalism, and above all because these two aspects of human presence on this planet—preservation of nature and prosperity for all—no longer seem totally incompatible. Nature is our ally; it should no longer be used as a dumping ground.

It is now finally possible to reconcile economic activity with the sustainability of our environment, both near and far. Not by reducing the former but by transforming the latter. And it's undoubtedly this way of thinking that has led me, organically, to look for ways to stop harming nature while pursuing the ideal of living as well as possible. I couldn't accept the idea that in order to improve people's lives, we had to destroy their environment. It was a kind of impossible pairing, and that's why I wanted to face it head on.

To be clear, far from having a consumerist mentality, I'm well aware of the problems associated with sobriety, but at the same time I can't get used to the idea that a part of humanity is being denied a decent life. Restoring hope, increasing purchasing power, fighting for jobs and against job insecurity—it all seems like squaring the circle. And yet I believe it's possible, starting with a change in energy policy.

In May 2019, I decided to leave the top management of a Suez subsidiary, Degrémont, the world leader in drinking water production, desalination and the treatment of wastewater, to become an entrepreneur.

At the time, it had been four years since the United Nations released its 2030 Agenda with its 17 Sustainable Development Goals. With more free time on my hands, I was able to immerse myself in analyzing each of them. I quickly came to the conclusion that the energy imperative was their common element: There was no point in hoping to achieve any of these goals without the utilization of energy, even if the intensity of that utilization was different for each of them. Any careful observer might have come to the same conclusion.

In this way, a personal conviction was combined with a vision inspired by this "concert of nations," which clearly reinforced my resolve, if ever such reinforcement was needed. I wrote a book called *Oui, c'est encore possible* (Yes, it's still possible), in which I expressed my conviction that energy is a kind of gateway, a necessary condition for achieving the 17 Sustainable Development Goals.

I didn't understand this right away. During my travels abroad to such countries as Malaysia, Singapore, Brazil and Morocco, I observed firsthand the injustices endured by certain populations, but I lacked the resources to challenge them. But by learning about them over time, in different latitudes, I arrived at a very straightforward conclusion, which can be divided into two parts: Water is a prerequisite for food; to have access to it, you have to expend energy.

With energy you can do anything. Without it, you can't do anything. The most important thing is to know what kind of energy you're using: physical force (human or animal), coal, gasoline, wind, sun or any of the nuclear energy sources that exist today.

Why I wrote this book

Life is a succession of events, more or less prompted, more or less enriching, more or less stimulating. By chance, some friends persuaded me to meet the Dalai Lama and his former prime minister, Samdhong Rinpoche. I went to Dharamshala and I remember telling them what was actively growing in my mind. I didn't just *want* to act; I was *determined* to do so. But there's a big gap between "I want to do something" and "I've decided to do something." Sometimes a simple word can make all the difference. After listening to me with a benevolent smile, the Dalai Lama simply said, "Go."

When you start a business, you're faced with a number of tasks. To help explain the different operations you have to perform, economists present the activity in a very simple way. With resources—money—you have to acquire several types of goods: raw materials, tools, a place to work, employees, a working method. It's like the art of cooking. To cook a dish, you need materials (meat, eggs, fish, vegetables, etc.), utensils (a knife, bowls), appliances (pots, an oven...), a place (a kitchen), skilled labor (a chef) and a method (recipes). You can't just make chocolate mousse out of the blue!

In all of this, two elements are crucial: money and a method. In general, the second is less problematic than the first, but without money, nothing can happen.

The year 2020 was spent finalizing the method—articulating the constituent elements and the business plan—and looking for funding. This is standard practice for any entrepreneur. Of course, a business plan is rarely perfect at the outset. It always has to be adjusted as the work progresses. As for money, it's also rare to have a great deal at your disposal. I was able to start with my own money and the help of friends, which was obviously not enough for Naarea's needs. So I had to appeal to investors for funds. At the same time, contacts with a wide range of people (entrepreneurs, politicians, scientists, researchers, intellectuals, journalists, etc.) allowed me to share my ambitions with others, hoping to attract the most qualified and knowledgeable people. The result was much better than I expected. Each time I explained the logic of the project (verbally or with a slide presentation, depending on the person I was talking to), I had to insist that it was possible to reconcile respect for nature with economic development. This either provoked skepticism or maybe a shrug of the shoulders or support, the intensity of which varied between approval, a desire to join the team—in however small a capacity—or a willingness to invest. Skepticism or rejection almost always stemmed from an a priori assumption: "if it were possible, people would know about it" or "it's too good to be true." It's an age-old reaction, and I wouldn't be surprised if the invention of the fork or the wheel triggered the same kind of reflex.

The hope attached to the project clearly fulfilled unspoken expectations, which in turn motivated growing support. Thus, in the fall of 2020, we set up our first strategic advisory board, made up of about ten people, including a former minister, a general, a company director, a consultant, a mathematician, a political scientist, an economist, a writer... Thanks to a banker friend, we were lucky enough to meet the first donor to Naarea's capital, which allowed us to really get started.

Recruitment of the team began in late 2021 and continued at a rapid pace. The initial group of about dozen people, all volunteers, grew rapidly. By the summer of 2023, the company had nearly 150 employees. This was an unusual occurrence for a startup that had been founded only 18 months earlier. Many young people wanted to join us, and some of the top experts in the nuclear field decided to get involved—because this is not only an industrial enterprise but above all a human adventure, a journey of knowledge sharing between the youngest and the most experienced, in the joy of contributing to a field where the only hierarchy is that of intelligence and skills, and where the common denominator is the desire to participate in an exciting and rewarding project.

The year 2023, which had not yet come to an end when I wrote these lines, was a year of accelerated development for the company, within a particularly dynamic and powerful French nuclear ecosystem. Two moments stand out: The project was a winner of the France 2030 program, which promotes technological development by funding targeted research, and it was selected as part of the French Tech 2030 program, which supports companies tackling major societal challenges through innovation.

What made such an adventure possible? Such questions are always difficult to answer. The temptation is to reconstruct the past by giving it a predetermined direction, but while it's conceivable to project oneself into the future, and more precisely into the future we hope for and work towards, it's virtually impossible to imagine all the possible paths that will lead there. This is because the human mind is never at rest. It develops new materials, discovers new uses for them and invents new processes that change the way we function without disrupting it too much, but it is capable of much more:

radical innovation by changing a habit, an approach, a line of reasoning. It's not easy to challenge long-held assumptions about reality, and yet it's something we should be constantly preparing for it.

Preparing for it means accepting that such disruptions are likely. Preparing for it means thinking outside the box. Preparing for it means putting a positive coefficient on uncertainty rather than seeing it as a threat. Preparing for it also means assuming that nothing is ever final, that it's always possible to think differently and change your habits. Preparing for it means thinking *into* the future rather than *about* the future. Clearly, none of us can claim to be adequately prepared for the unexpected, but we can all learn how to avoid getting stuck in the obvious and adopt a more open-minded approach.

Like most people, I believe that any revolution should lead people not to fight against their own innovations, however astonishing they may be, but on the contrary to strive to harness them for the common good. I don't know what upheavals artificial intelligence, for example, will bring, but I'm willing to bet—as is the case today with small nuclear reactors—that it can take us higher and further toward a better world, if we have the wisdom to see it as an opportunity and use it to our advantage.

Naarea's future seems assured, but there's more to it than that. It's now possible to make the entire nuclear industry sustainable. This is not just a question for one sector, but a development that offers great hope. The first task facing us today is to do everything in our power to stop climate change and the disasters it will bring with it. I have no doubt that we will achieve this goal by combining all the means at our disposal, starting with the one on which the others largely depend: the production of clean energy from nuclear technology.

We can only hope that this pioneering spirit will be emulated and that the world will resolutely enter a new era.